CORNISH ENGINES & THE MEN WHO HANDLED THEM

J.H. TROUNSON
M.B.E., A.C.S.M., C.Eng., M.I.M.M.

First published in 1967

by

The Royal Institution of Cornwall
Journal Volume V New Series, Part 3, 1967, p.213

Re-issued as individual work 1985

ISBN 0 90 4040 26 7

The Trevithick Society is very grateful to Mr J.H. Trounson and to the Council of the Royal Institution of Cornwall for permission to re-issue this lecture as a separate booklet.

Published by the Trevithick Society for Industrial Archaeology in Cornwall. Reg. Charity No. 246586.

Printed by Dowrick Design & Print Ltd., St. Ives, Cornwall.

FOREWORD

The Trevithick Society is fortunate in having as President John H. (Jack) Trounson, a man whose recollections of the steam engines which worked on the Cornish mines is only equalled by his knowledge of metal mining itself. Because he also possesses a rare gift of communication, anything he says or writes on the subject of Cornish engines is both instructive and entertaining.

When he held his audience spellbound at the Royal Institution of Cornwall in December 1966 his words were fortunately put on tape. The result, printed in the *R.I.C. Journal,* became a classic in the annals of Cornish engineering folklore. The Trevithick Society's decision to reprint the lecture in a booklet is to be welcomed for bringing his incredible-but-true stories to the attention of a wider audience.

Not only is the reprint being published in the Society's Golden Jubilee year, it also happens to be Jack Trounson's 80th, and 30 years after Cornwall's last large beam engine, Robinson's 80 inch, stopped work after a life of 101 years. It is therefore a fitting tribute to both Man and his marvellous Machines.

Kenneth Brown

CORNISH ENGINES AND THE MEN WHO HANDLED THEM

by

John H. Trounson

(Hon. Curator, Cornish Engines Preservation Society)

There is a wealth of information available concerning Cornish pumping engines but there is little in writing about the men who erected and worked them. As a step towards remedying the deficiency the Council of the Institution invited Mr. John Trounson, an outstanding authority on the subject, to give an informal talk on Cornish Engines and Engine Men. The talk was given at the Winter Meeting, 10*th December,* 1966, *it was tape recorded and the following, slightly edited, gives it in full; inevitably it lacks the charm of Mr. Trounson's delivery.*

I have been asked to tell you, quite informally, some of the stories that I have heard, mostly amusing but some tragic, and, having had the good fortune to see two large Cornish Engines erected, I will, possibly, be able to tell you a few things that you may not read in any book.

There being now a good promise of important new developments in Cornish mining, it may be opportune and interesting to look back at what was an essential part of mining machinery in the past—the Cornish Beam Engine.

Those who see a great Cornish pumping engine never cease to wonder how the beam, or *bob* as we call it in Cornwall, was got up into place.

The house in which the engine works is a tall three-storeyed building, each storey of which is known as a *chamber*. The front wall of the house, known alternatively as the *bob wall* or *bob end*, which carries the whole weight of the massive rocking beam, is made of great thickness to provide the necessary strength; in the case of a large engine it may be as much as six or seven feet of solid masonry—and built of granite if possible.

The steam cylinder, which may be anything up to 90 or 100 inches in diameter, with a piston stroke of 10 or 11 feet, stands vertically in the bottom chamber of the house, its upper end protruding through the floor above into the middle chamber. The cylinder rests on a great mass of masonry which forms the lower part of the house, weighing possibly 1,000 tons, and is securely bolted to it by massive bolts upwards of 20 feet in length. The piston works up and down inside the cylinder and the top end of the piston rod is connected to the beam, up in the top chamber, via the parallel motion gear in the middle chamber.

The outer end of the beam projects out over the mine shaft and to its end or *nose* the *main rod* is connected. The latter consists of massive baulks of timber, possibly as much as 18 or 20 inches square and 70 feet or more in length. These are joined end to end by long iron plates bolted to them and pass right down through the shaft which, in a deep mine, may be 2,000 feet or more. Attached to the rods at vertical intervals of from 200 to 300 feet are the actual pump rams or *poles* which raise the water in a series of *lifts,* each pump delivering to the next one above. The water is raised by the weight of the heavy rods descending after each up stroke, the purpose of the beam engine on surface being to raise the rods which in the case of a large pump may entail lifting anything up to 90 tons per stroke. It is because of the great weight that these engines have to lift that everything has to be made so strong and hence the size and weight of the bobs and the walls that carry them.

Before attempting to describe how the engine is erected and the beam lifted—the most spectacular part of the job—I had better give you a few of the Cornish terms for the different parts of the engine. As I have said, the beam is always known as the *bob* in Cornwall, each end of the bob is its *nose* and the great pivot on which the whole thing rocks is the *gudgeon pin.* The part of the bob within the house is the *indoor* end and that hanging over the shaft is the *outdoor* end. The bob consists of two huge castings joined together by suitable distance pieces and at both ends each casting is joined to the other by the *nose pin.*

The first step in erecting the engine is to get the bob up into place but before that can be done the lifting appliances must be positioned. The side walls of the house, known as the *wing walls,* are made thinner at the top, thus providing a substantial ledge on which to rest large timbers spanning the width of the house. For a heavy job these timbers will be at least 18 inches square. There will be two sets of them, one positioned above the centre of the bob wall and the other over the centre of the cylinder. Each set will consist of four timbers placed two and two, with one timber resting on top of the other—thus there will be two very strong beams, say 18 inches wide and 36 inches deep. The two beams will be placed about two feet apart with short strong timbers bridging their centres. From these latter timbers are hung, by heavy chains, the massive pulley blocks used for the lifting.

The big hand winches are then assembled at ground level and positioned around the house so that the wire ropes (chains in the olden days) from the winches to the pulley blocks pass through the window openings in the walls, or in through the front of the house over the bob wall which is still open and not yet boarded in. The winches are bolted down to a framework of timber loaded with many tons of big stones so as to prevent any possibility of them moving under the strain.

Anybody who has seen a Cornish Engine house will have noted that there is a very large doorway at the back of the house. This is known as the *cylinder opening* and it is through this that the bob, cylinder and other heavy parts are brought into the building. The bottom of this doorway will probably be some feet above ground level and the next step is to construct an inclined plane of heavy timbers or *skid road* up which the large parts of the engine can be drawn into the house.

All this preparatory work having been completed, which will occupy a dozen or so men about a fortnight or three weeks, everything is ready for erection to commence. First of all the great cast iron plate, on which the bearings of the bob stand, is placed in position on the top of the bob wall. Each of the two bearings consists of two parts, the lower portion is the *chair* and the upper, or actual bearing, the *stool.* The chairs are put in place but the stools are arranged to stand by their sides, resting on greased plates, so that they can be pushed into place by screw jacks quickly.

All this having been done, the bob is next hauled into the house up the inclined skid road. This is done by hand winches and powerful hydraulic jacks aided, possibly, by the winding rope of a steam traction engine. This part of the work may take a few days to accomplish but when it has been completed the bob will be inside the house and resting on the cylinder foundation with its outdoor nose almost touching the inside face of the bob wall and its indoor nose probably still protruding through the big doorway by which it entered the building.

The falls from the great pulley blocks over the centre of the bob wall are now attached to the outdoor nose pin and likewise the falls from the blocks over the cylinder centre are attached to the gudgeon pin. Everything is now ready for the lift to begin. This takes place in two stages; in the first the front winches work faster than the back ones and the bob is raised up to a steeply inclined position with its outdoor nose resting on the top corner of the bob wall and the indoor nose resting on the cylinder foundation or *loading.* The falls from the cylinder centre blocks are now transferred from gudgeon pin to indoor nose pin and those over the bob wall from the outdoor nose pin to the gudgeon pin. The second stage of the lift then takes place and when it has been completed the bob is in position and the bearing stools are pushed into place by the jacks, the winches are slackened off and the gudgeon pin rests in its bearings.

I first saw all this done in 1922 when the 90 inch engine was erected at the New Cook's Kitchen Shaft at South Crofty Mine. The bob of that engine was estimated to weigh about 48 tons and the foreman in charge of the work—the real brains on the job—was old Billy Jenkin of Chacewater. Billy had travelled all over the world doing this sort of work and driving engines in India, South Africa and at the various stations of the Metropolitan Water Board. He was an old man then; he always wore corduroy trousers with front pockets like those of breeches, into which his hands were usually stuck. On his head was a battered old trilby and protruding from the heavy grey moustache a little short clay pipe—generally upside down. Slow and quiet of speech, his bright and alert blue-grey eyes spoke volumes! If I remember aright the old chap rode to and 'fro daily from Chacewater to South Crofty by donkey shay—one of the last men to use the Cornish miner's conventional means of transport.

I was only a school-boy at the time but I suppose that I showed so much interest in the job that the men looked upon me as a sort of mascot and I was soon dubbed " I. K. Brunel " which degenerated into "Ikey ".

On the fine July morning on which the South Crofty bob was to be lifted I came in early with the men so as not to miss anything. There was a long ladder reaching from the bottom chamber of the engine house up to the top of the bob wall. As a signal mark of favour, Billy Jenkin said to me " You come up with me my son! " We climbed to the top of the wall where the old man took up his station for the great lift. Turning to me he said " I've stood in this place to welcome more bobs home than any other man living, and I don't suppose I'll do another " . . . he didn't, he died only a few months later.

Standing there, we could look down upon the winches in the bright sunshine of that summer morning. The men were stripping off their coats and even taking off their shirts in preparation for what they knew would be hard work. I suppose there were about two dozen men on the winches and a few mine officials in the yard but very few people compared with the crowd that would gather today to see such a sight. Standing with us on the bob wall was Jack Bryant, the principal rigger, a tall thin man who used to sail on Sir Thomas Lipton's great racing yachts. He was responsible for most of the lashing and tying of chains and ropes.

When everything was ready, old Billy Jenkin removed his pipe from his mouth and leaning forward, called down to his men " Are 'ee ready men? " Up came the reply " Yes, Mr. Jenkin ". " Right then boays, winnd up, the front winches a bit faster than the back ones ".

At first the winches turned easily, merely taking up the slack, but the " click, click, click " of the pawls gradually slowed down as the strain was taken by the men. Close as we were to the lifting tackle and the big timbers spanning the house we could now hear everything creaking and it made one realise what a weight was being lifted. Suddenly old Billy remarked to Jack Bryant " Jack, boay, that chain idden't right ". " No, he i'nt Mister, 'ees twisted ". Billy called down to the men on the winches " Stop, boays ". Jack got in amongst the chains and flogged the offending links with a hand hammer until he had got the twist out of them. Had this not been done until the full weight came on them it could have caused a dangerous shock if they had untwisted by themselves. Having surveyed everything and found it to their liking the two men decided to proceed and Billy again ordered " Winnd up ".

It was at about five or ten minutes past eight that somebody called out that they could see light between the bottom of the bob and the cylinder foundation on which she had been resting and then we knew that the lift was on in earnest—the struggling men on the winches had no doubt about it! By about quarter past ten the bob was up in the half-way position and while the men on the winches were having a well-earned " cup tay " the others were re-lashing the bob in the manner which I have already described preparatory to the second and final lift. For half an hour or so one could hear the ripple and roll of the chains being threaded through the eyes of the big lifting blocks and then, when everything was ready, the men on the winches resumed their labours.

Hardly a word was spoken between Billy Jenkin and Jack Bryant; they kept a close eye on all that was happening but they had confidence in their tackle. Slowly, very slowly, the nose of the great thing crept up over the bob wall as the men sweated and struggled in the sun and the slow clicking of the winch pawls continued. Now the great bob was sliding out over the shaft and began to look as she would do when working. At last Billy ordered " Stop, men ". Soon the screw jacks were at work pushing the bearings into place and, when all was right, there came the order to the men on the winches " Streak ", meaning " lower ". As the winches slackened off the great gudgeon pin sank slowly into the thick oil in the bearings with a sort of sucking sound. It was one o'clock and the job was completed though she had only left the ground about ten past eight that morning !

A year or so later this job was repeated at the East Pool and Agar Mines when the 90 inch engine from the Carn Brea Mines was re-erected at the then new Taylor Shaft at Agar. In this case the beam was even heavier, 52½ tons as compared with 48, but when lifting such a weight an additional 4½ tons is of little consequence. At East Pool the job occupied two days whereas at South Crofty it had been done during a Saturday morning. There was great rivalry between the two mines and I am afraid that the South Crofty people were strutting around with their chests up in the air !

However, to return to the erection of the South Crofty engine. After the bob had been lifted, the next thing was to get the cylinder bottom in its place. This weighed about 9 tons and, resting on rollers, was being drawn up the skid road by the winding rope of a powerful traction engine when the chain to which the rope was attached suddenly broke. Released from strain the wire rope flashed back like a serpent and had any of the men been standing near they would have been seriously injured or killed. It was an alarming incident and a warning to stand well clear of wire ropes when under strain.

With the cylinder bottom in place, the cylinder in its outer steam jacket or *case* (weighing about 25 tons all told) was drawn up into the house on a cradle and lifted, with the same tackle that had lifted the bob, into its place on the cylinder bottom. Then, with long plumb-lines hung from the indoor and outdoor nose pins of the bob, there followed the delicate operation of getting bob and cylinder correctly aligned in relation to one another. Provision is made in the design of the engines for adjustment of the main parts and with the heavy lifting tackle still in place it was necessary to lift both bob and cylinder slightly so that their relative positions could be adjusted by the aid of powerful jacks, after which everything was finally tightened down in place.

The way in which the alignment is done is illustrated by the drawing. As a bob goes up and down its nose pin describes an arc, A B C. If a perpendicular line is drawn from A to C, the horizontal distance B D is termed the *vibration.* If the vibration is halved, B E is the same length as E D. The perpendicular line E F is then the correct centre line of the piston rod and cylinder. This equalises the side strain on the swinging links which connect the piston rod *cap* with the nose pin of the bob. The main rod in the shaft is aligned under the other end of the bob in exactly the same manner.

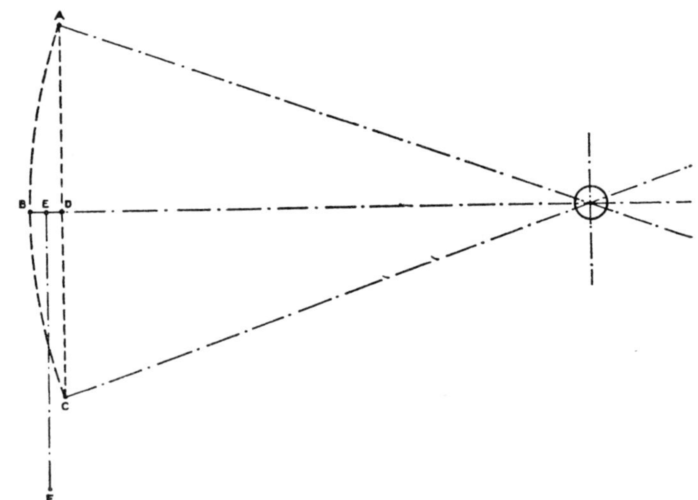

Method of Alignment

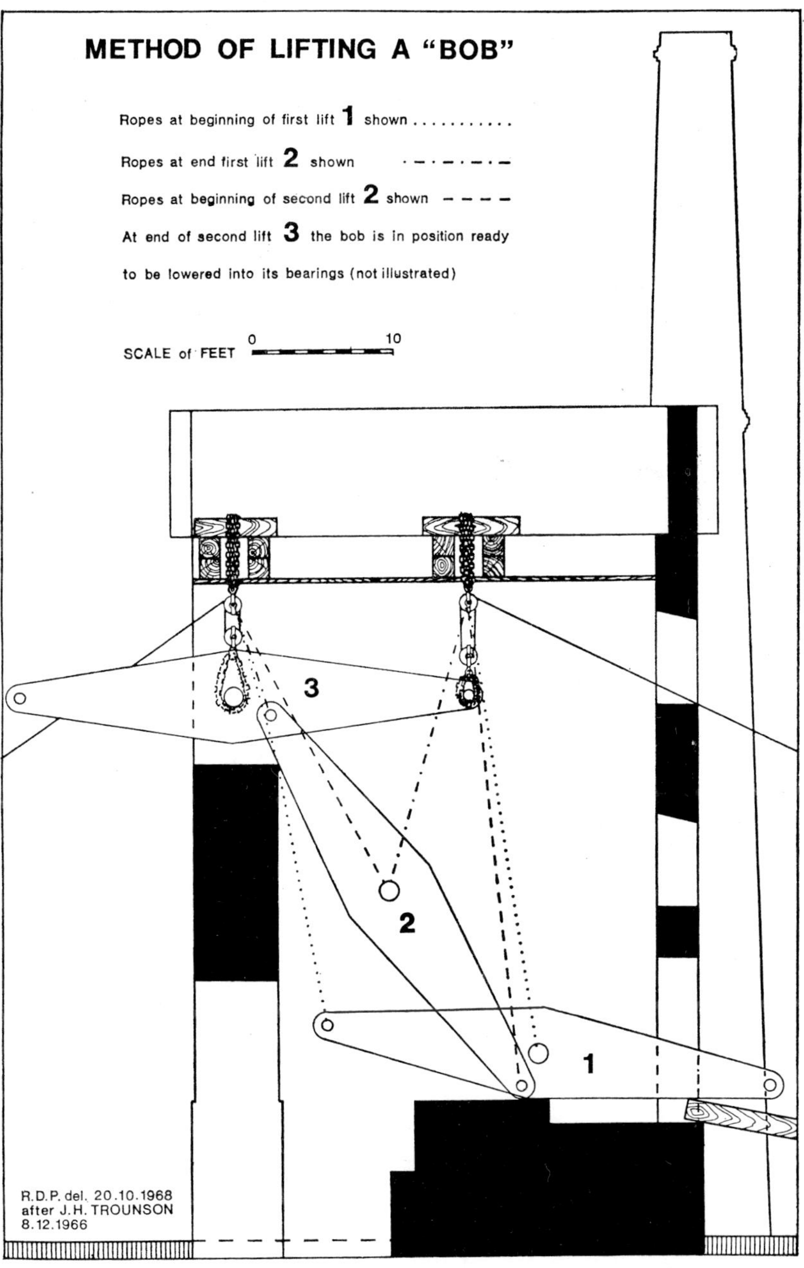
METHOD OF LIFTING A "BOB"
Ropes at beginning of first lift 1 shown ....
Ropes at end first lift 2 shown · — · — · —
Ropes at beginning of second lift 2 shown — — — —
At end of second lift 3 the bob is in position ready
to be lowered into its bearings (not illustrated)
SCALE of FEET
0
10
3
2
1
R.D.P. del. 20.10.1968
after J.H. TROUNSON
8.12.1966

The other heavy parts of the engine could then be assembled and while that was in progress the tedious process of making the rust joints could proceed. In the early days of heavy engineering there were not the machine tools available for facing off the joint flanges of large castings such as those encountered in the Cornish engine. James Watt therefore evolved an iron cement that could be hammered into the joints and which would set almost as hard as the cast iron itself. We do not know what ingredients he used and, consequently, today it is left to each erector to use his own preferred mixture; it is usually made up of iron filings, sulphur, sal-ammoniac, etc., all made up into a paste with red and white lead putty. Special tools are inserted into the joint and placed a few inches apart, the space between being filled with the jointing mixture which is then hammered in with a blunt-ended caulking tool until it literally rings like a piece of metal. The retaining tools are then withdrawn and inserted again a few inches away from the piece of joint already made and the process is repeated until at last the whole joint has been completed. In the case of a very big joint like the bottom end of a 90 inch cylinder it will take one or two pairs of men many days to go all the way around the cylinder.

When the 90 inch engine I am talking about was removed from Wheal Grenville and re-erected at South Crofty, work on the dismantling of the engine and excavations for the new engine house started almost simultaneously; this was about the 4th January, 1922. To economise in time and money the house was built in mass concrete, with a certain amount of reinforcement, instead of in the conventional stonework. Apart from the red brick upper part of the stack, it was estimated that 7,500 tons of material went into that house.

During the building of the house preparations were being made in the shaft to receive nearly 1,200 feet of very large Cornish pitwork. In addition to cutting massive hitches in solid rock in which the supporting girders would be placed, a big chamber was blasted out about 600 feet from surface in which a balance bob or counterpoise could be installed. As soon as the house was roofed the re-erection of the engine at its new site commenced and simultaneously the pitwork was being lowered down the shaft and the new boilers were being set by the side of the engine house and built into place. During the whole of this work only a few night shifts were worked on surface, these, I believe, being during the making of the rust joints. As I have said, the job commenced early in January that year but practically everything was finished by the time that the engine made her first trial run on or about the 4th of November.

I was away at school at the time and greatly regretted that I was unable to see her put to work. I heard later that an amusing thing occurred on this occasion. She made her first stroke quite normally, but when they tried to bring her *indoors* again she refused to move. All but one of those present were deeply puzzled but he, a rough old dog named Ernie Williams, could see what was wrong. At last Ernie said " If all the lot of you will get out of the house I'll bring her in all right ". " Well, what is it ? " they replied. The only answer they got was " Get out of the house or I shan't do it ". In the end everybody had to agree to go outside and Ernie immediately picked up a spanner and adjusted the condenser injection valve. He saw that the valve wasn't opening and, consequently, the steam was not being condensed so that there was steam beneath the piston as well as on top and the engine couldn't move. Ernie confided to me later, " I could see what 'twas, I picked up a spanner and adjusted the valve and the engine come in like a bird ". No doubt several people were very, very annoyed ! Ernie Williams long ago joined the Great Majority and so I can tell you the story, but it only shows how experienced men can be fooled by a stupid little detail.

There are many other amusing stories that could be told about the erection of these engines. At the East Wheal Rose lead mine, near Newlyn East, when the tremendous new 100 inch engine was erected in the '80's, Matthew Loam, who was the consulting engineer, decided that he was going to put her to work himself. However, it appears that when fitting the woodwork of the front of the house around the bob, the carpenters had not allowed sufficient room for its movement. The consequence was that as Loam brought her indoors for the first time there was a rending crash and splintering wood flew in all directions and everybody rushed out of the house !

One of the most interesting stories I have heard was in connection with the erection of a replacement engine at Old Sump Shaft at the Carn Brea Mines in 1887. The old 76 inch cylinder engine there was completely worn out and was such a wreck that it was decided to replace her with a completely new 80 inch engine. However, for a number of reasons, it was not feasible to build a new house and the new engine had to be erected in the existing building. It was therefore essential that the change-over should be made as quickly as possible to minimise the flooding of the deeper workings. The job was therefore postponed until August month when the inflowing water was at its minimum.

The man in charge of the operation was the great engine erector Tom Rowe (pronounced " Rawe ", and frequently described as " 'ee with the beard " !) As the old engine had to continue working to the very last moment little could be done to make the necessary alterations to the house (such as the cylinder foundation) until the old engine had been thrown out. Rowe, however, made his plans very skilfully and the job was done in record time for an engine of this size.

The parts of the new engine were distributed at strategic places around the engine house and shaft, the new beam facing the bob wall and ready to be lifted in *over the shaft*. The plan was that, as soon as the old cylinder and other parts inside the house had been got out of the way, two steeply inclined skid roads of heavy timber were to be constructed with the top end of each resting on the bob wall, one inside the house and the other outside, and therefore bridging the shaft. The new bob was already standing on rollers and as soon as the old engine could be dismantled the bearings of the old bob were to be removed and that too placed on rollers. The two bobs would then be linked together by long chains and, as the old bob descended the inside incline, so its weight, acting through the chains, would help to pull the new bob up the outside incline.

The scheme worked perfectly—to the chagrin of the manager of the mine ! He had wanted to see the operation but he was not popular and Tom Rowe decided that the job wasn't going to be done while he was there. Having asked Tom when he was going to do the lift, and having been assured that it wouldn't be that evening, the said gentleman got into his carriage and was driven home to Truro. The moment his back was turned Tom began to move very quickly and when the manager arrived next morning he was flabbergasted to see the new bob up in place.

The story was told me by an old engine driver called Luther Martin. As a child he was taken by his mother up to Carn Brea to see the strange sight. There were hundreds of sight-seers from Pool and all around the area. It was a fine star-lit night, but to enable Tom Rowe to check on the movement of each bob and to be sure that each was moving in the desired path he had them illuminated with hundreds of miner's candles stuck on with lumps of clay such as miners used to carry their lights on their hats. One can imagine that scene under the velvet midnight sky, the men working away on the winches, the click of the pawls, the great candle-lit mass of cast iron slowly moving up the incline, the hundreds of people standing there silently watching and " 'ee with the beard " confidently in charge of the whole operation and probably having little to say. Luther Martin told me that he would never forget what he saw that night.

The next urgent task was to pull out the old cylinder loading and Mr. Barton has recorded that this entailed building in 270 tons of new masonry. While all this work was taking place on surface, the shaftmen were replacing a considerable amount of the pitwork with new. Of course, in those days it was possible to employ as many skilled men as could be crowded in on to the job—and men were prepared to work in those days ! Nevertheless, it was a remarkable feat that the whole job was completed and the new engine went to work within seventeen days.

It is said that the building and erection of Hocking's 85 inch engine at the United Mines in Gwennap was an all-time record. From the time of placing the order at Perran Foundry, the engine was manufactured, the engine house built, the pumps installed and the engine erected in, I believe, exactly sixteen weeks. It is almost incredible to us now how such a thing could be done, but our forefathers were not afraid to work and knew how to use their brains.

A very quick piece of erection was performed at Levant Mine. The little 45 inch engine there, which was scrapped after the mine closed in 1930, broke her beam and Harvey's cast a new one. Of course, it was only a small thing weighing perhaps 14 tons but still quite a lump of metal. It was delivered on site by Harvey's big traction engine which was fitted with a powerful crane and driven by the celebrated but cantankerous old Joe Beck.

When Joe arrived at the precipitous slope at the foot of which was the engine house, he got off his engine to size up the situation and then said to the mine engineer " Shall I put 'un up with the engine for 'ee Mister ? " I don't think the offer was accepted, but Joe was quite prepared to swing that bob in on the spot !

Another interesting job was at the Grenville Mines in 1906 where the 80 inch engine on Goold's Shaft broke her beam. It happened at a time when there was frost. It has been noticed that metal loses its strength at low temperatures as well as high. A friend of mine who was in the Royal Engineers during the First World War told me that as the Germans were retreating they were destroying the railway tracks behind them by blasting the rails and, to enable the British Armies to follow up as quickly as possible, a single line had to be constructed from the debris of two. They found that when it was necessary to cut a rail in the very cold weather prevailing all they had to do was to put four nicks in it with a chisel, lay it across another rail and give it a heavy blow with an 18 pound sledge and it would " break off like a carrot ".

The day before the bob broke at Grenville, the working engineer came in to see the stoker at Fortescue's 90 inch engine boiler house (the engine which was later re-erected at South Crofty) and he happened to remark that they had two nice pumping engines there at Grenville. On the following morning, at about 6 o-clock, Goold's bob broke right in two just as the engine commenced her stroke. The piston came down with such force that the cylinder cover and bottom were both smashed and even the lower part of the cylinder jacket or case was cracked.

By the time that the poor harassed working engineer came around again on his daily tour of inspection the other engine had been speeded up in the forlorn hope of preventing the mine from being flooded. When he entered Fortescue's boiler house he was greeted by the same young stoker with the quite needlessly irritating remark " Well, Mister, you've got two nice engines now ! " He was nearly fired on the spot and deserved to have been sacked for saying such a thing.

Holman's cast a new and heavy bob for Goold's engine and there is a photograph of one side of this leaving their works in Camborne, hauled by a great team of horses. I have another photograph showing how the two sides of the bob were lifted into place, one at a time. To save removing the cylinder from the house the bob was taken in over the shaft, the weight being carried by heavy timbers bridging the gap between engine house and headgear.

At East Pool Mine they had an old 70 inch engine which rejoiced in the nickname of ' Old Kitty '. To facilitate raising and lowering heavy pitwork parts in the shaft it was decided to erect a powerful steam capstan and a big sheer-legs standing above the shaft. Because of the shape of the headgear this *sheers* could not be of the normal two-legged type and it consisted of a single very large pitch-pine leg which supported one end of the wheel beams, or *caps*, the other end of which rested on the existing headgear. The caps were surmounted by a single large and heavy cast iron sheave wheel. The whole thing resembled a man with one leg amputated at the thigh and it proved a most awkward and out-of-balance thing to erect. Old Mr. John Tonkin, the Engineer of the mine, was superintending the lifting and things were not going too well. Under such circumstances Mr. Tonkin could be ' difficult ' and this was certainly not the time to crack a joke at his expense. Being " crowst " or lunch time, the men in the crusher house had walked around to see how the job was proceeding.

One of them remarked "I think I can see a quicker way to get that great leg up, Mr. Tonkin ". " What's that my man ? " (irritably) " Plant the seed and let the b—grow ! "

It is said that old Tonkin nearly jumped in the shaft with annoyance.

Before leaving the subject of engine erection may I tell you a little story about the engine house down on Porthtowan Beach which once had a beautiful red brick top to its stack but this was taken down after it was converted to a summer dwelling house. This forms a part of Wheal Lushington and I believe the house was built to accommodate an old second-hand rotary beam engine brought over from Ireland.

Billy Jenkin, the foreman on the erection of the 90 inch engine at South Crofty, was quite a young man in those days but he was engaged to erect this engine. He told me the story. " Mister, I was told to go down and see what gear was wanted to put her in, so I had a pony and trap and went down to Porthtowan, and there she was lying in the meadow near the engine house. I sized she up and decided what winches and gear I needed, but in the evening a clerk came out from Redruth and said ' we arn't going put the engine in after all ', but he paid me for my day's work ".

Apparently there was some shady work going on in connection with the financing of this company and, as a result of some of the shareholders looking into matters, everything was stopped to a moment's notice. This was not the only engine house that was built but never had an engine put in it !

About 1899 or 1900 an unsuccessful attempt was made to work the massive dumps in the Gwennap area in the mistaken belief that they contained sufficient tin to make this profitable. The necessary plant was erected on the old United Mines and consisted of a 34 inch Cornish stamps engine bought from Killifreth, with 120 head of stamps (the last big Cornish steam stamps to be erected) and an 80 inch engine to pump the water for dressing the ore. Had the venture prospered it was the intention to unwater the mines and the 80 inch engine would then have formed the nucleus of the pumping plant required. Consequently, the starting of this engine was a matter of considerable local interest and several people were in the house to see her make her first move. Captain Joe Odgers of Carharrack was to put her to work and he described the scene to me graphically. He remarked " See, when you are doing that you have got to put your best hand forth ! " He was a tall old man and I can imagine how he must have braced himself up on the great day as he walked forward and took hold of her handles, determined to make a good showing of ' putting her to work '.

Incidentally, this engine was designed by Matthew Loam and built by Harvey's in 1868 for Batter's Shaft at the West Chiverton lead mine near Zelah. She is said to have been Loam's favourite engine and after West Chiverton closed down Harvey's bought her back and put her in store as she was too good to break up. When the Gwennap venture started she was re-erected in Garland's engine house which had originally contained an 85 inch engine. She only worked a few months there and was ultimately sold to Condurrow United Mines and erected in the new engine house (still standing) near Beacon, Camborne. She was broken up about 1916 and narrowly missed being used again. After the disastrous smash of Pascoe's 80 inch engine at the Basset Mines that Company would gladly have purchased the Condurrow 80, but she had been broken up only a few months before.

Speaking of putting an engine to work and ' putting one's best hand forth ', I witnessed a pleasing example of this at the Metropolitan Water Board's Kew Bridge Pumping Station in London in 1946. The Board had six magnificent Cornish engines there, five of which have been preserved including the largest which was built by Harvey's of Hayle and which has a cylinder 100 inches in diameter with a piston stroke of 11 feet.

On the occasion of which I am speaking a large company of engineers had been invited to see the last formal working of the great engine though, actually, she worked again once or twice for other engineering bodies to see. However, on this great occasion the driver who was in charge of her was a ' Cousin Jack ' from Troon, appropriately named Harvey. He had spent most of his working life driving these engines in London and now he was to put the 100 to work for, possibly, the last time, and before a large body of engineers, many of whom had come up from Cornwall. The old chap nearly burst with pride as he was called on to put her to work. Attired in a clean boiler suit, he walked forward and took her handles and she went off like a bird. It made one realise what 'putting your best hand forth' meant!

I have already mentioned what happened when the 100 inch engine started at East Wheal Rose. I had a most interesting description of this too from old Captain Tom Piper who, when he was well advanced in years, was a night boss at Tincroft Mine. As a boy he worked at East Wheal Rose and saw both the 90 (which later worked at Wheal Agar) and the 100 go to work. The 90 was the first to be started and on that occasion Tom Piper was told off "to hold the heads of the gentry's horses". In describing the 100 he told me that the house was lined with oak wainscotting and that the place was "kept very select"!

Those of you who have read Mr. Barton's excellent book on the Cornish engine will know all about the three monster engines that were built in Cornwall for draining the Haarlem Mere in Holland. These truly amazing engines had an 84 inch high-pressure cylinder standing in the middle of a 144 inch low-pressure one. The centre piston had a single piston rod but the outer or annular piston had four rods, all being coupled to a single massive crosshead from which radiated a number of bobs. Fortunately, one of these, the Cruquis Engine, has been preserved by the Dutch Government.

It is not generally realised, however, that the second largest Cornish beam engine ever built was one of conventional type with a single cylinder and single beam. This had a cylinder 112 inches in diameter though the piston stroke was only 10 feet which was rather short for an engine of such immense size. This was constructed by Harveys for the old Southwark and Vauxhall Water Company which later formed a part of the Metropolitan Water Board. It stood where Battersea Power Station now stands and commenced work in October 1858.

I knew one of the men who drove this great engine, the late Stephen Polglaze of Marazion, who gave me a lot of information about her. Stephen was a small man, I wouldn't say he was an inch over five feet, and light in weight with it. He was emphasising the fact to me that the gearwork of the big engine was beautifully balanced and, notwithstanding her great size, she was an easy thing to handle. He remarked " Look, Mister, I arn't a very big man, but I could master she all right ". I have known very much smaller engines which required all the weight and strength of a six foot man to press down the bottom handle, so if five feet of Stephen Polglaze could 'master' the 112 she must indeed have been nicely balanced !

The last pumping engine to work at Wheal Peevor was a 70 inch designed by the celebrated Samuel Grose. She had stood derelict and rusting for many years at Wheal Johnny, north-west of Camborne, after that mine was abandoned. Incidentally, this is the shaft where a boy fell down and lost his life a few years ago which triggered off the agitation for the better fencing of mine shafts throughout the county.

A very peculiar thing happened in connection with this 70 while she stood disused at Wheal Johnny, but before telling you about this I had better explain one or two technicalities.

In common with most Cornish pumping engines this one was *levered.* This means that the stroke of the piston was greater than that of the rods in the shaft. This is achieved by making the cylinder end of the bob longer than the shaft end thus giving the engine a certain amount of " leverage " over her load. In a Cornish engine each end of the bob is usually made 20 inches in length per foot of stroke and, as the Wheal Johnny engine had shaft and piston strokes of 9 and 10 feet respectively, the bob was 20 inches longer indoors than out, and was thus considerably heavier at the piston end. In addition to this there was the weight of the piston and its rod, the parallel motion and all the other moving parts indoors. When the outdoor nose of the bob of an engine is disconnected it is always necessary to support the indoor end with timber or else the whole thing will be completely unbalanced and the engine will come in with a crash.

On the other hand, it is the legally recognized custom in Cornwall that the top piece of shaft rod, with its heavy iron plates, is a part of the engine itself and is always sold with it. Consequently, when the pitwork was removed from the shaft at Wheal Johnny the top rod was left hanging from the nose of the bob. It must have been a particularly large and heavy rod for it almost exactly balanced the weight of the engine parts indoors. However, after a lot of rain the timber became water-soaked and heavier and the engine would then go 'outdoors' of her own accord. After a long spell of dry weather the rod would dry out and lose weight and then the engine would make an 'indoor' stroke! I never met anybody who actually saw her move but for a number of years the engine performed these ghostly movements all by herself like those old-fashioned toys in which little Mistress Sunshine or little Mister Rain come out or go into their house. The local people used to say "Johnny's made a stroke again, the weather's changing!"

This was probably the only engine that ever moved herself in this way but it had a very beneficial effect. Instead of the cylinder becoming rusty inside, as might otherwise have happened, when she was dismantled to take her to Peevor about 1912, they found her "as bright as a new shilling" inside. Externally, however, the poor old engine had become terribly rusty while she lay there and, in addition, thieves had stolen a lot of her bearing brasses. A man using explosives to get the brasses out of her accidentally blew some of his fingers off one Sunday morning.

When removed to Wheal Peevor much of the gearwork was in such a state that the only thing that could be done to it was to set it up in a lathe and turn it again. This was done by an old retired Naval Petty Officer who, though a genius, could only work when half drunk. The old lathe that he had to use was a very rough tool too and, consequently, it was not surprising that the engine was "nothing much to look at".

Our old friend Billy Jenkin comes into the story again at this point for he was also the foreman on this job. When the engine was ready to start Billy decided that he would put her to work himself. Mr. Cyril Stewart of Treleigh told me that he was a boy at the time and, wanting to see the engine make her first stroke and not wanting to be chased out of the house by Billy Jenkin, he crept upstairs when nobody was looking and hid himself in the top chamber. Unfortunately, some timber had been left under the bob and had been overlooked in the final preparations. As she made her first stroke there was an almighty crash and young Stewart was so scared that he nearly jumped from top to bottom of the engine house in one leap !

Those of you who are familiar with a Cornish engine will know that, attached to the indoor nose of the bob, there are two upright members bridged at the top by a heavy cross piece of rectangular section. This is called the *catch wing* and its purpose is to prevent the piston going too far and striking the bottom of the cylinder. Normally, as the bob comes indoors the catch wing comes down to within about an inch of the *spring beams* provided to arrest it, but if steam pressure is rising in the boilers (it tends to vary according to the state of the fires) it will come further and touch the beams. If the driver is not alert and does not close the steam valve slightly the wing will strike the spring beams with increasing force at each succeeding stroke.

I have said that this engine looked rough externally but she also had very rough handling at Wheal Peevor. One day a visitor went into the house and found an old man driving her who was obviously as deaf as the proverbial post. She started to touch her beams but the old man didn't appear to realise it and at last the blows became so violent that loose bricks on the top of the stack were hurtling through the slates on the roof. The old man suddenly asked his visitor " Is she touching, Mister ? " Yes, the poor old engine had a very rough life there !

While speaking of the handling of engines I remember other incidents that were related to me by various drivers. One of these was an elderly man called Hocking who drove Sara's 65 inch engine at Wheal Kitty, the last Cornish engine to work in the St. Agnes parish. In his younger days old Hocking had driven an old-fashioned beam winding engine, with " plug " type valve gear, at the Herodsfoot lead mine near Liskeard. These must have been tricky things to handle and he described to me how, at night, they kept an old horn lantern in the " plug door " (the doorway beneath the bob) to illuminate the crank and so be able to see its position when swinging the engine to get it over ' dead centre ' at the commencement of a wind. Modern winding engine drivers, accustomed to beautifully controllable machines with interlocking safety devices, would almost certainly have an overwind with such an engine on the very first wind !

Sometimes in the olden days if they had a heavy weight such as a complete bucket or " drawing lift " to raise a few feet in the shaft, they would do it by attaching the load to the main rod with heavy chains and then bring the engine indoors slowly thus lifting the weight perhaps 9 or 10 feet. Old Tom Magor told me that this was being done at Killifreth one day when he was driving the 80 inch engine. They told him to bring her indoors but she wouldn't move and though he gave her more steam she still would not move. Little did they realise it but a flange on one of the big castings they were trying to lift had jammed somewhere in the shaft and the thing was virtually immovable. Old Tom told me " They said ' go on Tom, give her steam ' and I did so until there was a crash—the bob had broken on one side ! " At that time of severe depression the mine was in financial difficulties and could not possibly afford a new bob, or even one new side. The broken one therefore had to be repaired and strengthened by fitting it with a " kingpost " and " bridle-rods ".

Those who saw the old beam winding engine at Levant, which worked up to the time that the mine was abandoned in 1930, will remember the famous old character, Tom Taskis, who drove her. Earlier in life he had driven the big horizontal steam winder which Messrs. Holmans built for the Botallack Mines in 1907.

As usual with engines of this type, one winding drum was keyed to the crankshaft so that it was bound to rotate with it but the other drum was a loose one free to rotate on the shaft when the clutch was disengaged. A clutch drum is normal practice on a winding engine so that the relative positions of the two cages or skips in the shaft may be varied to suit whichever level the engine is hoisting from. In this case, however, the type of clutch fitted was a friction one instead of the normal jaw clutch. These patent friction clutches were not a success and in spite of repeated adjustments the Botallack one was apt to slip at times and this could have had disastrous consequences if it had got out of control when hoisting men.

This is just what happened one day when Tom was driving. I believe there were men in the cage which was coming up while the other one was going down empty. Suddenly, the drum of the ascending cage began to run backwards but with wonderful presence of mind Tom stopped the engine, reversed her instantly and turned on steam again and as she ran in the same direction as the slipping drum the faulty clutch just managed to 'bite', thus bringing the falling cage slowly to a standstill before it hit the bottom. It had been a near thing and, but for Tom's superb enginemanship, there would have been a disaster. That man ought to have been decorated but, needless to say, he wasn't!

As a matter of interest, after Botallack closed down this winder was re-erected at the then new Taylor Shaft of the East Pool and Agar Mines in about 1923. Rebuilt with a new jaw clutch, it continued to give splendid service until that mine too closed in 1945. It still remains in the house though it is now in a partially dismantled state.

Old Tom Magor, whom I have already mentioned in connection with the breakage of the bob at Killifreth, told me that, when a little boy, he was standing at the top of the United Mines looking across country towards Baldhu when he saw a great cloud of steam go up at Nangiles Mine. Realising that there had been a boiler explosion and, boy-like, being unable to resist the temptation to see what had happened, he walked over there immediately. He told me that the thing which remained in his memory was seeing a basin full of blood. They were rough old days and boiler explosions were regrettably frequent then.

Strangely enough, it was one of Tom Magor's own sons

who was nearly involved in what could have been another disastrous explosion at a much later date.

Just before the First World War, Wheal Coates, on the cliffs at Chapel Porth, was being reopened. The mine was being drained by a Cornish pump driven by a horizontal tandem, compound, rotary steam engine, the steam for which was generated in a Lancashire boiler. Through shortage of fresh water in the area they were having to use water pumped from the mine for the boiler, but this water was not only very red and dirty but highly saline because the workings extend beneath the sea. The result was that, though they did not realise it, a great accumulation of salt and mineral matter was forming on the water side of the furnace tubes. As a deposit of this nature is a very effective heat insulator, the steel plates were becoming badly over-heated.

The younger Magor told me that it was towards the end of a day shift that he went into the boiler house to clean the fires, i.e. to clean out the clinker, and to get everything ready for the afternoon shift driver who would be following him. As he opened the fire doors he was startled to see that the crowns of the two furnaces were coming down like inverted bowler hats of glowing red-hot metal. Nobody could have blamed him if he had run for his life but with amazing courage this little hunchbacked man stood his ground, closed the dampers, lifted the safety valve, put on the feed pump to cool the boiler down and raked out the fires. At enormous risk that man had prevented an explosion and the boiler was subsequently repaired, but I doubt if he was ever thanked for what he had done.

That leads me to mention a boiler explosion that did take place but it had a humorous instead of a tragic ending. This happened at the boiler house of the big 90 inch Agar pumping engine at about the turn of the century. There were then five large Cornish boilers there in a line, No. 1 being the one nearest to the engine house. This boiler had been cleansed and then put to work again but it had unwisely been brought into steam too quickly. A big boiler should be allowed several hours in which to generate steam otherwise very severe strains are apt to be set up in the steel plates through expansion. In this case, not only had the boiler been unduly rushed but the boiler fitters had not done their job properly and they had not detected that the lower fitting of the water gauge was blocked with scale. Consequently, the glass was still showing plenty of water when in actual fact the water level was dangerously low; this was the primary cause of the explosion which followed.

On this particular evening the stoker came down the stairs from the engine house carrying the kettle which he wished to boil so as to make tea. He opened the firedoor of No. 1 boiler and, with the aid of the long poker or rake, placed the kettle on the bridge at the back of the fire. A few minutes later he again opened the same fire door with the intention of lifting out the kettle and was frightened to see the crown of the tube then coming down. He ran for his life and had just reached the open door at the other end of the boiler house when the tremendous explosion occurred. Apparently he was not hit by the steam but the compression of air flung him several yards and he fell into a pile of soft coal standing outside the building and, though not seriously hurt, was stunned for a while.

Though the other boilers were not seriously damaged the building was almost wrecked and the front wall was blown under the balance box, at that moment in the up position. This effectively stopped the engine working immediately, and now everybody was rushing around in the dim light of candles and oil lamps looking for the body of the deceased stoker. Suddenly they heard a sound and, turning around, saw a figure with a pair of white eyes in a black face staggering into the wrecked building—several people nearly had heart attacks !

A boiler explosion is always treated as a very serious matter and the inevitable Board of Trade inquiry followed. The Inspector asked the stoker to tell the Court what had happened leading up to the explosion. He accordingly rehearsed the story up to the point where he had again opened the firedoor and saw the crown coming down. " What steps did you take? " asked the Inspector. Quick as a flash came the reply " b— long ones ". The Court dissolved in laughter !

One of the last and most interesting sights I witnessed in connection with a Cornish engine was the occasion when the cylinder cover joint of the 90 at Taylor's Shaft at East Pool blew out and had to be re-made. This is done with a piece of plumber's lead pipe, beaten flat, and coated with red lead putty. The weight of the cover, about 7 tons, would be sufficient to press the lead flat but, of course, the dozens of big nuts around the periphery of the cover have to be screwed tight and must be tightened evenly all around in order to make a good joint.

As I walked up the stairs into the middle chamber I wished that I had a camera or that there had been an artist present to capture the scene in colour. Surrounding the top of the cylinder were about a dozen men, mostly elderly, wearing all sorts of old battered hats and caps, working away with big spanners, two men to a spanner, like a great rowing team. Hardly a word was being spoken, only the dull sound of the spanners on the nuts and the shuffling of feet on the wooden floor.

These great engines were strangely picturesque in themselves, but the scenes of human endeavour often associated with them formed pictures which those who witnessed these things will never forget.

It fell to my lot to design the last Cornish pump to be installed in Cornwall—perhaps in all the world. This was for the Castle-an-Dinas wolfram mine, near St. Columb, in 1942/3. One would like to be able to record that the motive power was a Cornish beam engine but at a time of acute war-time difficulties this was not to be and a diesel engine and reduction gearing had to be used instead. The actual pumping plant together with the twin balance bobs was installed by the late Walter Langford of Lanner, a genius if ever there was one. He was then well over 70 and the shaft was extremely wet and unpleasant for working in, but it didn't seem to deter him in the least. On one occasion the weather was very cold and rough and as he had to work on surface that day I felt concerned that a man of that age should be there working under such conditions. When I remarked on the weather he replied in a loud voice " Well my son, 'tis breakage weather, 'tis breakage weather ! "

I thought afterwards how true the old man's remark was. Time and time again I have noticed that when there has been a breakage on a Cornish mine, such as the failure of a main rod of a pumping engine, the weather has often been abnormally cold or wet thus making the work all the harder for the men. Robinson's 80 inch engine at South Crofty was the very last Cornish pumping engine to work on a Cornish mine. On the occasion when the last new rod had to be put in there the weather was bitterly cold. I have a photograph showing the new rod laid out on the ground near the engine house with the carpenters and shaftmen drilling the *pin* holes in it, but the cold was so severe that tarpaulin screens had been put up to shelter the men from the east wind and braziers were burning by the side to try and keep the men warm. The photo shows snow on the ground and icicles five feet long hanging from the roof of the " dry " or miner's change house.

Talking of famous old personalities on the mines, Jimmy Pryor, who drove Robinson's engine, was certainly one of them. Incidentally, his son, a very spruce little figure, who had driven some of the largest winding engines in Cornwall, died not long ago. He was a remarkable man for his age and was much older than he looked.

However, to return to the father. As a youngster I was sitting by his side on the settle in Robinson's engine house late one evening and I was so fascinated by his tales of the past that I lost all count of time and it was about half past twelve when I left. On my return home I found that people were out looking for me as it was feared that I might have fallen down a shaft.

Amongst his accomplishments Jimmy Pryor was a skilled clock maker. It is said that at Comford pub one evening, for a wager of a pint or two of beer, Jimmy consented to be blindfolded and the parts of a grandfather clock which he had not seen were laid before him but he succeeded in assembling the parts and had the clock working within, I believe, half an hour.

He was a small man with a slight stoop, a thin grey moustache and little goatee beard and with a slight cast in one eye. In the engine house he wore an old waistcoat that was so oily that it seemed to be polished like a boot, and on his head he wore a dark blue or black skullcap akin to the white hatcaps traditionally worn by the miners beneath their hard hats. When he left work he was always attired in a bowler hat and long dark overcoat, with his hands stuck in the side pockets. He walked daily in all weathers from his home at Lanner to South Crofty and back again at the end of the shift. On occasion the Engineer of the mine has met him on the road and stopped his car and offered him a lift but the reply he got was "No thank you Sir, I prefer to walk". He was an independent old character; during the 1914-18 War when the Unions first became strong in Cornwall they tried to force everyone to join but he refused to do so. The Committee said that they would stay in the engine house until he did join. "Well", said Jimmy, "the place is nice and comfortable, you had better have cup o'tay, make yourself at home". They then walked with him to and from work for days, but it made no difference—Jimmy wasn't joining any trade union!

I remember another evening talking to him in the engine house when the Pitman, who rejoiced in the unofficial name of 'Joe Pigs', came in to say that they were going down to change one of the "clacks", or water valves. He made arrangements with Jimmy as to the bell signals that should be rung from underground in order that the engine might be stopped and then moved as they required. Before Joe Pigs left the house Jimmy asked him which clack he intended changing, and when he told him Jimmy remarked quietly that that was the wrong one. With colourful language 'Pigs' asked him what the hell he knew about it. Jimmy snorted and replied "Well, go down and find out". The amusing thing was that Jimmy was proved right within half an hour. He had so got the "feel" of that grand old engine that he could tell exactly where there was any trouble in the pitwork.

Another 'character' but of a very different type was big Ernie Williams who has already been mentioned in connection with the starting of the 90 at South Crofty. The last engine he drove was the other 90 at the Taylor's Shaft at East Pool. Ernie told me that, when a boy, he was working with a tributer on a dump at Wheal Sisters at Lelant. What I am about to relate happened, I believe, on the last day of March, 1886. The sky became very overcast and suddenly there was a terrific flash of lightning accompanied by a tremendous clap of thunder. Ernie said that he felt as if someone had struck him in the back with a 'dag' (an axe) and the man he was with cried out. He involuntarily ran down over the side of the dump and as he did so he caught sight of the stack of Wheal Mary pumping engine falling in a cloud of dust.

Though both Ernie and his partner had received shocks, neither was seriously injured but there were some near escapes from death when the stack was struck, and that was only a quarter of a mile from where they were working. Apparently the settle in the engine house was, most unusually, placed at the back of the cylinder instead of in front where the man can watch the gearwork in operation. A moment before the lightning struck the driver had risen from the settle and walked to the plug doorway and shouted up to the "lander" in the headgear. That saved his life for only a second or two later the settle was buried in tons of rubble as the stack crashed through the roof and two floors. Incidentally, the lightning stroke travelled right down the main pump rods in the shaft and, passing in through one of the deep levels, struck two tributers at work there and gave them quite a serious shock.

This story ends with a rather peculiar coincidence. When the stack fell a large granite quoin struck one of the boilers and made a large dent in the plate. Many years later when Ernie Williams was working at East Pool, a traction engine arrived hauling a second-hand boiler that had been purchased by the Company. When he looked at it, there was a big dent on the top at the back end!

After the death of the late Edgar Trestrail of Redruth I had the privilege of going through his mining and engineering papers and there found a most interesting letter written by the famous Matthew Loam to his father, Nicholas Trestrail, in the year 1896. This describes how, in about 1826, he, Loam, saw experiments made at the Consolidated Mines at Gwennap by the famous (and irascible) Arthur Woolf and his own equally famous father, the great Michael Loam. Incidentally, we are deeply indebted to Mr. T. R. Harris for his recent splendid life of Arthur Woolf.

Matthew Loam's letter is so interesting that it is worth reading a part of it to you this afternoon. He wrote:—

> " Perhaps you will excuse here a circumstance which I remember as a schoolboy of Woolf versus myself. I had heard he was about to experiment on the expansion of steam at Consols, and I had a holiday to go out to witness it. Before the experiment was finished they retired to discuss and my Father strictly charged me not to meddle with the apparatus; and especially the glass gauges. Somehow I approached them and broke one and the rush of steam frightened me and soon brought my Father and Woolf to the scene; and I hid myself. Father at once told him that I had broken the glass. He asked " Where is he ? " and I had to appear. He was in a gentle mood and looking at me quietly he asked " Did you break the glass " I replied " I did Sir and I am sorry for it ". He at once took out 6d. from his pocket and gave me saying **" here is 6d. for telling the truth ".** This must be nearly 70 years or more, since I suspect it was the first 6d. that I ever obtained, but I have never forgotten it ".

I remember going out to the United Mines one evening with the late W. J. Opie of Redruth who must have been much older than he looked for he had actually worked there as an engine boy in his youth. It was extremely interesting to hear him describing, on the spot, what the mines had looked like when they were working, and where the various engines stood —even their houses are now nearly all gone.

He worked at Garland's engine. This, I believe, was an 85 inch but, as I have already mentioned, an 80 was erected in that same house at the turn of the century. Old Opie described to me how, on occasion, things would be going well, the fires burning nicely with a good head of steam and he would be out sitting on the grassy bank by the side of the engine house when, suddenly, the shaftmen would come up from underground to say that they had to change a clack. He, as the boy, would be sent running down to Taylor's or Hocking's or Cardozo's engines to tell them to " crease up ", meaning increase speed as Garland's was about to stop for half an hour or more. The water at the United was so ' quick ' that when an engine stopped for repairs the bottom level would be quickly flooded unless the other engines increased speed immediately.

I believe that Mr. Opie told me that he was working there as an engine boy when he was only nine years old. In those days the engine drivers had to fire their own boilers instead of having a stoker as in latter days. However, in the case of the large engines they were provided with a boy to help. His duty was to wheel out the ashes and clinker from the boiler house and to help the driver generally about the place. Mr. Opie told me that, as a child, he was terrified walking home in the dark past the trees at Trevince. While still very young he went to work underground at South Penstruthal Mine. His job was to wheel stuff in a wheelbarrow but the levels were so narrow that in order to protect one's knuckles it was necessary to wrap them with ' bal shag '—the coarse woollen material used for making joints in pitwork and for innumerable other purposes on the mines. He said that one could clearly see the grooves which generations of men's hands had rubbed in the soft rock at the sides of the levels in that mine.

While in reminiscent vein I cannot resist telling you a story or two about the well-known engineer, Mr. Francis W. Michell of Redruth, as related to me by his son, the late Mr. W. A. Michell. The father, Francis W., had a considerable consulting practice and was mechanical engineer to several mines. He was a short man, who usually wore a silk hat, and was a very great character !

One day he was at Wheal Uny and was entering the door of a boiler house where the working engineer had foolishly just installed a water pipe running low across the top of the doorway. The silk hat hit the pipe with such force that the brim was split and the topper driven down over his eyes. As its owner struggled to remove the remains from his head he positively hissed through his teeth " Take that b— pipe out "!

West Seton was another of Mr. Michell's mines and one day he received an urgent message to go there at once as there had been a boiler explosion. Taking his son, William, with him he left immediately by pony trap but when they got to the bottom of Tuckingmill they were met by another man from the mine who said " Mr. Michell, there's a boiler gone up at the mine ". " Yes, yes I know " replied Mr. Michell irritably, " No Sir, this is another one " the man said.

Needless to say, by the time that Mr. Michell reached the mine he was not in a very good mood. He first went to the site of the second explosion and, by the time that he had finished his investigations there, some hours had elapsed since the first explosion and the remains of that boiler and what was left of the surrounding brickwork had cooled off. When the inspecting party arrived at the scene of the first explosion they were horrified to see the legs and feet of a man just appearing from beneath the wrecked boiler. Thinking that it was a corpse, somebody jumped down into the lower brick flue to investigate, only to find that it was an old tramp who had thought that this was a nice warm place so he had lain down to sleep. Mr. Michell, thoroughly peppery by this time, was furious, " Arrest him ", " Send for the Police " he shouted—there was hell to play !

In the old days when a mine stopped she was said to be ' scat '. If at all possible, they would strip out all the pitwork and raise it to surface before the water rose. Of necessity this had to be done quickly and it was hard and dirty work. It was traditional that the men doing it should be provided with new underground suits made of the usual white duck with lined jackets to keep out the wet and cold. I recently came across an estimate made in 1896 for stripping Killifreth pitwork. One figure astounded me, the cost of providing 15 men with new suits worked out at 30/- per man ! Heaven alone knows what such suits would cost now.

There are a few other miscellaneous stories which I may just as well tell you today.

When the rather grandiose 90 inch engine was erected at the Highburrow East Shaft of the Carn Brea Mines in 1893 it is probable that the Company wished to provide itself with ample reserve pumping power in case the pumping dispute between the neighbouring East Pool and Agar Companies were to lead to a critical water situation in the area. This did in fact come to pass a few years later, but the new Carn Brea plant was always far more powerful than was needed by the mine itself. The real reason for such a large new engine was probably not fully realised, locally, and it was thought by some to be mere ostentation. I remember an old engine driver saying to me sarcastically "When Captain White (the Manager) and 'Nicky' Trestrail (the Engineer) put up that engine they thought that they were going to frighten the world!"

Little Stephen Polglaze whom I have already mentioned in connection with the great 112 at Battersea told me a rather interesting story about the big engine. They had an outstanding fitter there who could file two pieces of metal so truly flat that if placed together they could not be separated by hand. His skill was called into play when a hidden defect appeared in the piston rod of the 112. Apparently there was a cavity at one side of the rod which was covered over by a thin skin of metal but this had worn through in rubbing against the packing in the stuffing box. This fitter cut a piece out of the rod and dovetailed in a replacement block of metal and, after caulking over the edges and cleaning it up with files, the repair could not even be detected.

Speaking of repairs, the old 90 inch engine at Wheal Agar had been grossly over-loaded for years and was becoming very unreliable. When reading through the weekly reports for the period one sees that the breakage of a main rod in the shaft would be followed by trouble in the boiler house. Then something would burst in the shaft and the following week there might be trouble with the engine itself—the breakdowns were continuous.

One day the underground manager, Captain Bill Treloar, was taken ill with influenza and he was away from work for about a fortnight. Somebody from the mine called to enquire for him. Capt. Bill being anxious to know how things were progressing asked " How is Agar engine going ? " Back came the reply " 'ansome Capn., she havn't stopped once for the fortnight ". Capt. Bill commented " Well m'son, that's good hearing even if tin't true ! "

The famous 80 inch engine at Robinson's Shaft, South Crofty Mine, now happily preserved, has a most interesting history having worked at four different mines as Mr. Barton records in his book. When at Tregurtha Downs, near Marazion, she frequently ran at 13 strokes a minute in winter time, with very large pitwork too, namely, 20 inch. I have heard three or four men say that they have seen her doing 12 or 13 strokes a minute, one of them being old Mr. Joel Clemo of Redruth. He was originally a miner who later became a travelling draper (a 'Johnny go fortnight') and he told me that wherever he went by on the high road from Marazion he used to stop the pony, take out his watch and time St. Aubyn's engine (named after the shaft) to see how many strokes she was doing per minute.

Mr. Joseph Blight, for long the Engineer of South Crofty, (and a delightful character) told me that on one occasion he was staying with friends over the week-end at Marazion. On the Sunday it was raining heavily but, to use his own words, " In the morning I was a good boy and went to Chapel and then, in the afternoon, I set out to see that wonderful engine. I took my big umbrella and went to the top of the hill from where I could see her. Yes Mister, she was going for it all right, doing a full twelve and a half. I walked down to the mine and went into the engine house and took off my wet coat and sat down in the settle and timed her carefully ".

Those who remember the late Mr. Blight with his dark moustache and sidewhiskers, the bowler hat, the neat turned-down collar and black tie (as once worn by Naval Petty Officers) will be able to picture the scene. Mr. Blight, incidentally, worked at Harveys in his younger days and helped to make some of the parts of the six 70 inch Cornish engines that worked in the No. 1 House of the Severn Tunnel pumping plant.

The centenary of Robinson's engine was celebrated on the

11th September, 1954. I have a photograph showing the late Lord Falmouth, who was President of the Cornish Engines Preservation Society, and Mr. Van der Pol, the eminent Dutch engineer, placing laurel wreaths on the cylinder of the engine while her driver looks proudly on. Later that afternoon it fell to my lot to deliver a commemorative address before a large number of people gathered around the shaft while the grand old engine worked away above us.

It was on a Sunday, the 1st May, 1955, when she was in her 101st year, that she made her last stroke as the new electric pumps were then ready to take over from her. Of course, she had not actually worked for 101 years as she had stood idle for a while at each of the three previous mines until sold and erected elsewhere. Nevertheless, she probably did more work than any other Cornish engine that was ever built.

On the previous evening, the last day in April, several engineers gathered to have a last look at her working, amongst them was the late Mr. Tregoning Hooper to whom we are principally indebted for the formation of the Cornish Engines Preservation Society.

It was a beautiful evening with a glorious sunset which seemed to fit the occasion. Some of us went out in the field just south of the shaft where we could get a full view of the engine and its house and watch the sun setting on that great moving red bob for the last time. As the sun went down in a blaze of colour its light shone through the windows silhouetting the indoor end of the bob as it rose and fell. Then, as the light faded and the outdoor end of the moving bob grew faint in the shadow of the headgear, the lights came on in the bottom chamber and through the window by the driver's settle we could see the beautiful polished, tapered, handles of the gear lifting and dropping as the shining plug rods slid silently up and down. A thin wisp of smoke rose vertically from the stack into the evening sky and from the neighbourhood of the condenser could be heard a faint burbling and the sound of running water; there was no other sound.

As I watched I could not help ruminating on the changes which have taken place since that engine was built at Hayle in 1854. At that time Florence Nightingale had not yet commenced her great nursing work, the mails were still being carried across the oceans in sailing and paddle ships and there was as yet no railway linking Cornwall with the rest of England. Neither the internal combustion engine nor the pneumatic tyre had been invented and by night the Cornish engine houses were lit by candles! Needless to say, nothing was known of nuclear physics and there were no such things as wireless or television. What a vastly different (and more peaceful) world it was in which that old engine had commenced life, but she had lived through it all. One could not help feeling that it would be entirely wrong if such an historic and faithful old servant were to be broken up merely because more modern machinery was taking her place.

Before leaving the mine I returned to the engine house and sat in the settle with the driver and we talked now that the others had gone. The stoker came up from the boiler house for a moment and then went back again and we two were left with the grand old engine. There she was working away as if she had another hundred years of life ahead of her. The gentle sound of the steam expanding in her cylinder like the breathing of a great giant, the melodious hollow ringing note of the catches at each stroke and the slight thud and " swish " of the steam slides and the clamps kissing the steam horns and the handles. Everywhere, a gentle pleasant warmth, a sweet smell of oil, snow-white walls, clean floors and the table, with its top scrubbed white, in front of the old settle with, of course, the inevitable brown tea-pot and cups on the table.

Before I left I stopped her and then put her to work again—the last time I was likely to have an opportunity to do so. As I went out through the big glass cylinder doorway and walked down the steps my mind was full of memories.

On the following morning, the Sunday, several people congregated in the engine house to see the end. The talk was of old times and though we were all there to see her make her last stroke there was a strange and awesome atmosphere about it all. It was like waiting for the passing of a great monarch and we knew that it must come soon; this was to be the last Cornish engine to work on a Cornish mine.

During the hours of waiting the final tests were being made of the electric pumps and at last, at 1.05 p.m., the dread signal was rung from underground to stop the engine. Several elderly men came downstairs from the middle chamber and we stood watching silently as the driver stopped her and pinned her handles down for the last time—I believe he had tears in his eyes.

There was an awful silence, it was indeed like the death of a king. People drifted slowly and unhappily out of the engine house and the driver and stoker sat down in the settle as if they were lost and didn't know what to do with themselves. I couldn't help wondering whether the spirits of the great Samuel Grose who designed her and the men of Sandys, Vivian's foundry at Copperhouse who made her were hovering around us that day. This was, truly, the end of a mighty epoch in the long history of Cornish mining and engineering.

Through the generosity of the Directors of South Crofty Ltd. it later became possible for the Cornish Engines Preservation Society to preserve this historic engine, but had it not been for the initiative of W. Tregoning Hooper, W. A. Michell and A. Treve Holman the Society would not have existed. It was a happy inspiration when that body decided to make the preservation of this famous engine its memorial to the late Mr. Holman who had been its first Chairman.

Unfortunately, however, it has proved to be beyond the resources of a relatively small Society to maintain a number of these engines, and particularly their houses, indefinitely. It is therefore good news that, provided a sufficient dowry can be raised, the National Trust is prepared to take over this responsibility. Thus the future of the engines will be assured for all time.

The Cornish engine played a very great part in the development of both engineering and mining and I feel that some of this lore, in scattered fragments though it may be, should be recorded while it is still green in somebody's memory.

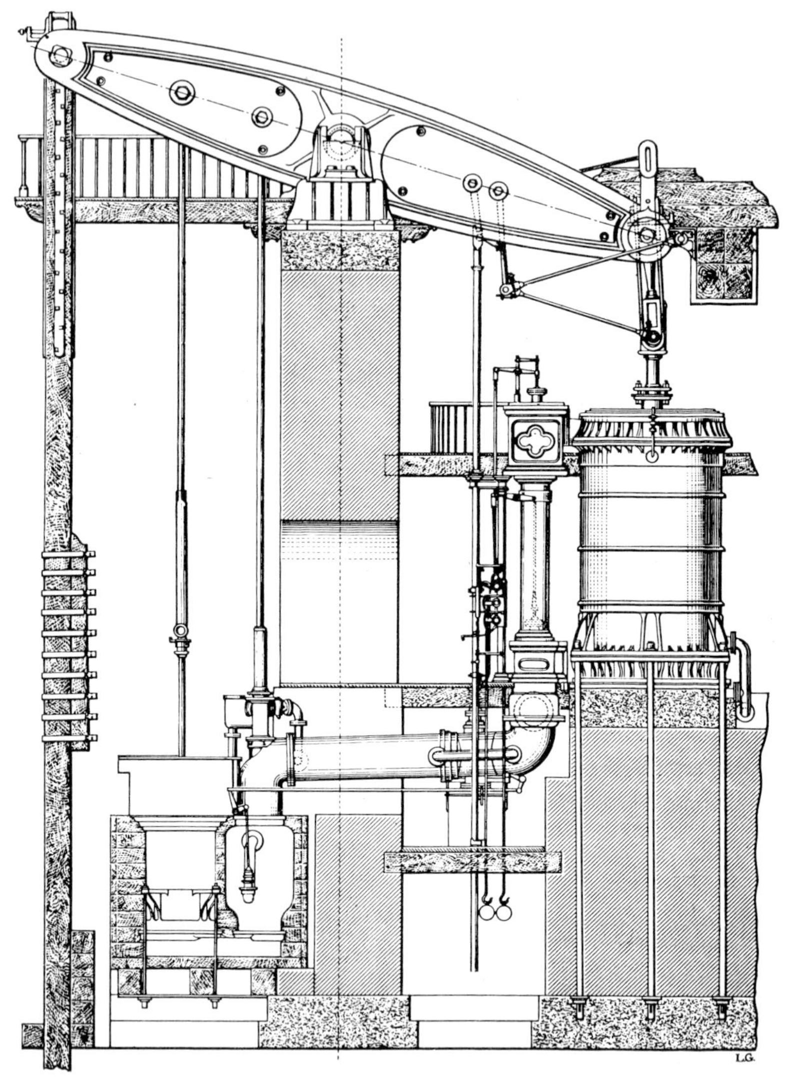

Diagram of Cornish Engine.